Spreadsheet Activities in Middle School Mathematics

Second Edition
Revised for Macintosh and ClarisWorks

John C. Russell
Natrona County School District
Casper, Wyoming

NATIONAL COUNCIL OF TEACHERS OF MATHEMATICS
Reston, Virginia

Copyright © 1992, 1998 by
THE NATIONAL COUNCIL OF TEACHERS OF MATHEMATICS, INC.
1906 Association Drive, Reston, VA 20191-1593
All rights reserved

Second edition

ISBN 0-87353-462-X

The publications of the National Council of Teachers of Mathematics present a variety of viewpoints. The views expressed or implied in this publication, unless otherwise noted, should not be interpreted as official positions of the Council.

Printed in the United States of America

CONTENTS

Preface v

Introduction 1

Instructional Models and Ideas 5

   Number Patterns 6
      ARITHMETIC SEQUENCES 6
      FIBONACCI NUMBERS 7
      PASCAL'S TRIANGLE 7
      3 × 3 MAGIC SQUARE 8
      4 × 4 MAGIC SQUARE 8
      MULTIPLES 8

   Basic Arithmetic 9
      DOING THOSE OPERATIONS 9
      AVERAGES 10
      PROPORTIONS 11
      RANDOM NO. GRID 11
      EXPANDED NOTATION 12
      DECIMAL EQUIVALENTS 13

   Geometry 14
      RECTANGLE AREA VS. PERIMETER 14
      PLAYGROUND DESIGN HELPER 14
      GENERATING PI 15

   Probability 16
      TEN-COIN EXPERIMENT 17
      FIVE-DICE EXPERIMENT 17
      URN WITH REPLACEMENT 18
      EIGHT-COIN BINOMIAL 19
      MATCHING BIRTHDAYS 19

Communication across the Curriculum 20
    MY STATE REPORT 20

Miscellaneous 21
    MESSAGE FROM SPACE 21
    WORD VALUE 22

Word Problems 23
    MANAGING MY MONEY 23

Graphing 24
    PARABOLA 24
    SOLVE LINEAR SYSTEM 25
    COORDINATE PLANE 25

PREFACE

In 1992 I compiled a set of spreadsheet activities based on the work of many middle school students and teachers in Casper (Natrona County), Wyoming. The spreadsheets were developed using Better Working Spreadsheet, an Apple II product of Spinnaker Software, Cambridge, Massachusetts, for which NCTM subsequently acquired distribution rights.

Now, the Apple II has faded considerably from the school scene and has been replaced by newer brands and models, including the Apple Macintosh. A number of powerful spreadsheets exist on the Macintosh platform. ClarisWorks, by Claris Corporation, is an excellent example of an integrated package containing more than adequate spreadsheet and charting power for use in the middle grades.

This book is an update of the original edition, aimed at users of ClarisWorks on the Macintosh. The spreadsheet examples have been written in ClarisWorks version 4.0. Although version 5.0 has arrived on the scene, these examples can be opened and used in either 4.0 or 5.0. They can be rewritten for other programs and platforms, too. Similarly, the classroom ideas should be easily adaptable.

INTRODUCTION

Since the writing of the first edition, a great deal is new in the application of technology to elementary school mathematics.

Consider the powerful new products on the market. Broderbund Software's Tabletop Jr. offers a rich environment in which young students can investigate classifying and sorting; be introduced to Venn diagrams and picture graphs; model real-world situations dealing with numbers, attributes, and time; and more. Turtle Math from LCSI takes Logo to new heights. No longer do text-based commands just dictate to the graphics environment: the mouse can now alter the graphics and send implications back to the text-based commands. Geometry, language, and creativity intertwine in exciting ways in this program. Number Connections, a product of Sunburst/Wings for Learning, provides the tools to explore numerals, number words, sets, number lines, base ten, and addition and subtraction in a laboratory designed for the comfort of our youngest students.

Houghton Mifflin and MECC are marketing a comprehensive set of hands-on environments in their MathKeys series. Whole numbers, probability, geometry, decimals, and fractions are among the grades K–6 mathematics topics addressed in this set of programs.

What these and many other programs bring to the classroom is the opportunity for students to investigate mathematics actively, developing models that can answer many questions and raise even more. Various spreadsheet programs deserve a place among these other excellent programs, since they provide an arena in which open-ended numerical questions can be effectively addressed.

Beyond the many exciting new software titles appearing in the '90s, the Internet with its principal subset, the World Wide Web, is far and away the single greatest drawing card in computing for most of us. Information on absolutely any subject is readily available, whether it be useful or frivolous, true or false, calming or controversial. Mathematics teachers and students can easily find support of all kinds using any of a large selection of search engines. Type in the keyword *spreadsheet,* and you will find literally thousands of pages dealing in some way with the subject. Just one example is "Spreadsheets in Education" (**www.smc.univie.ac.at/~neuwirth /spreaded/spreaded.html**), abounding in ideas, examples, resources, and links to related pages. We could quote many other addresses. The Web is a very fluid place with pages coming and going constantly, so it is best you go searching on your own.

New software and the Internet aside, it really always has been and continues to be important that students "think in the middle of a problem" and "think around the edges of a problem." Spreadsheet activity invites them into the "middle," to meet and solve problems head-on. Spreadsheets also let students linger "around the edges," observing patterns, investigating what-ifs, and asking questions that arise from initial answers. Today's spreadsheets further offer powerful graphing features that older versions didn't; pairing the dynamics of spreadsheet formulas with colorful pictures of the resulting data provides wonderful support for students both in the middle and around the edges.

As school districts revise their curricula, the wise ones are acknowledging that students learn best what they construct themselves. That acknowledgment often manifests itself in performance-based goals or outcomes, which call on students to question, collect, organize, interpret, conclude, and report. Spreadsheets can play a central role in such activities, and they reach beyond the mathematics classroom to science, social studies, and other subject areas.

Although the formal use of spreadsheets is probably appropriate beginning in the upper elementary school years, primary school classrooms can certainly offer readiness opportunities for the younger children. Lettered and numbered grids on the classroom floor with students giving commands for the movement of words and colored chips among cells can prepare them for their future use of labels, values, and formulas on computer screens.

Computerized spreadsheets can be used for instruction in a couple of ways. One use features teacher-presented, canned models. Those models, stored on drives and ready for retrieval by students, can be used to study different concepts. Students observe the displayed data resulting from their input and then see patterns, hone their prediction skills, and rise above traditional arithmetic mechanics.

Alternatively, students can be the producers of spreadsheets. They can introduce into a blank spreadsheet screen those words, numbers, and rules that organize that which they wish to communicate mathematically.

INTRODUCTION

In this package, like the earlier Apple II version, we will include canned examples with some ideas regarding their use. But if that is the extent of the use they get, it is hardly justifiable given the power of software now dedicated to geometry, probability, graphing, and the communication of ideas. Instead, more than before, it is important that teachers and students jump into these examples, applying their needs and reasoning to building new ideas into these spreadsheets and their related activities. Without such a contract, these spreadsheets pale in comparision to the competition.

In the earlier edition we gave a concise tutorial in the use of the Better Working Spreadsheet software. In the days of the Apple II there was no consistent interface among different programs from different publishers, and Better Working Spreadsheet was by no means a mainstream program. Having used one Apple II spreadsheet was no guarantee that principles and operational details were obvious in another spreadsheet. A positive outgrowth of the Macintosh operating system, however, is the predictable set of menus and mouse actions across virtually all software products. So, in this edition we will dispense with any spreadsheet tutorial, making the assumption that you are familiar with the Macintosh or will pick up the necessary how-tos on your own.

Thanks for your interest in helping students strengthen their mathematical thinking and technology skills through the use of spreadsheets.

INSTRUCTIONAL MODELS AND IDEAS

Here we begin our use of the spreadsheet in learning some mathematics. A list of models created toward that end follows, along with some ideas on how and why to use them.

We emphasize that the small library presented here is intended as a seed. Our examples are not comprehensive. There are not enough examples to teach, for instance, all there is to know about division, or introductory probability, or sixth-grade mathematics. We have a beginning, though, to use as a "convincer" that spreadsheets are exciting, dynamic tools with which to address certain topics in the mathematics curriculum.

There is not a hint of drill and practice here, nor does the spreadsheet serve as a tutor. What it does is show things—what arithmetic really does to numbers, how some numbers influence other numbers, and what large experimental samples look like. It is left for the teacher to take advantage of the benefits.

This work is addressed primarily to the teacher, not the student. By that we mean teachers can pick and choose from the collection any ideas that might work and reshape them in ways appropriate for their own classrooms. The descriptions and activities presented are not intended as textbook pages or worksheets.

We recommend a variety of teacher-student-computer configurations. Sometimes two students working on problem solving at each computer in the lab will be effective. Other times, the teacher should use a single computer with a large display to lead a discussion. There may be times when five- or six-student teams learning cooperatively at one computer in the classroom provide maximum effectiveness. The teacher should implement practices appropriate to the situations and the class.

We have categorized the sheets discussed here into areas in which they might be involved. Surely and naturally, there is overlap.

Number Patterns

Patterns and sequences are always a source of interest. It is fun to see what happens next and to appreciate the beauty brought about by certain rules. The spreadsheet can model such patterns for study with more ease and motivation than paper and pencil.

Open ARITHMETIC SEQUENCES from the Number Patterns folder.

Enter a "first number" in cell F5 and a "second number" in cell F6, and instantaneously you have 100 terms of any arithmetic sequence you wish. Cumulative totals are kept along the way. Negative numbers work just fine. Decimals work, but note that the sheet is set to display numbers generally, and therefore patterns can be obscured by that presentation. You might choose to unlock all cells (Edit/Select All, then Options/Unlock Cells) and then change numbers to a fixed number of decimal places (Format/Number).

Discuss the model ahead of time to assure students' understanding of arithmetic sequences, then give them access to this sheet.

Exercises

1. What is the sum of the first 100 positive whole numbers?
2. In the sequence for the problem above, what do you notice about consecutive pairs of cumulative totals (two odds, two evens)? Why?
3. What is the sum of the first 100 positive even whole numbers?
4. In the sequence for the problem above, how many odd cumulative totals are there? Why?
5. Look at the pattern of the final digits in the cumulative totals. Why are there no 4s or 8s?
6. Use 5 for the first term and 3 for the common difference. Look closely at the last digits of the cumulative totals from term 1 through term 10. Read them forward and read them backward. What do you notice? Any idea about what is happening?
7. Try different combinations of first term and common difference and look for patterns.
8. If I give you 10 cents the first day and increase my gift by 3 cents each day, how much will I give you on the 77th day? How much will I have given you altogether after 100 days?
9. Rich Uncle Frank wants to give you $1000 each week for 100 weeks. Because of temporary trouble with his bank, he can't come up with $1000 a week at the beginning. So he decides to start with less than $1000 a week and raise the donation by $5 each week. What number of dollars should he start with, allowing for a $5 raise each week, to bring the final total to exactly the same total as if he had given $1000 each week?
10. MAKE UP YOUR OWN EXERCISES!
11. Extension: Create your own spreadsheet model for another sequence, for instance, perfect squares—1, 4, 9, 16, and so forth.

INSTRUCTIONAL MODELS AND IDEAS

Open FIBONACCI NUMBERS from the Number Patterns folder.

Numbers like Fibonacci numbers, fabled in history and nature, are generated by providing two seed numbers. The third number is the sum of the first two, the fourth number is the sum of the second and third, and so on.

Exercises

1. If the first two numbers are 5 and 6, what are the next three numbers?
2. Does it make a difference whether a sequence starts with 1, 2 or 2, 1? If the first two numbers are unequal, which order gives the larger following numbers?
3. True or false? Doubling the first two numbers doubles all others.
4. Begin with two 7s. Divide any number down the list by 7, mentally. What happens? (If it divides evenly, the number must be a multiple of 7.) Change the first two numbers to 6s. Again, mentally divide any number down the list by 6. Do you notice the same phenomenon? Can you find an exception to this rule? Why does it work? (Consider the distributive principle here.)
5. MAKE UP YOUR OWN EXERCISES!
6. *Extension:* Create a spreadsheet like FIBONACCI except that the terms are generated by alternately adding and subtracting, for instance, 1, 1, 2, 1, 3, 2, 5, 3, 8, 5, 13, 8, … Look for patterns.

Open PASCAL'S TRIANGLE from the Number Patterns folder.

Pascal's triangle is somewhat of a two-dimensional version of the Fibonacci sequence. Beginning with a triangular shape of three 1's, the triangle is continued in a downward direction by placing 1's on the peripheries and filling in with the totals of the two entries immediately above.

Exercises

1. In the original Pascal's triangle, check the frequency of (1) evens and odds and (2) primes and composites.
2. Change the periphery numbers to evens exclusively, odds exclusively, or primes exclusively and examine the resulting numbers in the triangle. Can you come to any conclusions?
3. Have the class choose a large target number, say 67 465, and challenge the class to come as close to that total as possible near the middle of the eighth row, say cell J12, within a given time limit. Try this exercise several times, honing those estimation skills. Involve positives and negatives if appropriate to the grade level.
4. MAKE UP YOUR OWN EXERCISES!
5. *Extension:* Create a spreadsheet to generate triangular numbers: 1, 3, 6, 10, 15, 21, 28, …

Open 3 × 3 MAGIC SQUARE from the Number Patterns folder.

Studies in number patterns can take game formats. Although it is not a number pattern environment like the previous models, 3 × 3 MAGIC SQUARE has to do with the reasoned manipulation of numbers, so we'll put it here. A magic square is a square grid of numbers whose totals horizontally, vertically, and on the two major diagonals are equal. Magic squares can be of various dimensions; this one is only 3 × 3. Magic squares give students practice in addition and offer opportunities for the development of problem-solving strategies.

The task is to use the mouse or keyboard to rearrange the numbers 1 through 9 in the grid so that all eight totals around the edge are equal. The numbers 1 through 9 must be used once and only once in the grid.

Exercises

1. Solve the puzzle.
2. Are there other solutions?
3. Allow yourself the integers −4 through 4, if appropriate to your grade level. Can you form a magic square?
4. Can you make an arrangement, using 1 through 9 without duplicates, in which the sums are not all equal but rather form a sequence of consecutive whole numbers? How about with other numbers you choose?
5. MAKE UP YOUR OWN EXERCISES!
6. *Extension:* Create your own spreadsheet number game.

Open 4 × 4 MAGIC SQUARE from the Number Patterns folder.

Exercises

1. Solve the puzzle.
2. Can you solve the puzzle so that there are four overlapping 3 × 3 magic squares within the solved 4 × 4 magic square?
3. MAKE UP YOUR OWN EXERCISES!
4. *Extension:* Create a magic square with a larger dimension. Study magic squares; learn when there are solutions and when there are not; become your class's expert on magic squares.

Open MULTIPLES from the Number Patterns folder.

Cells E7 and H7 accept any whole numbers. Below each of those cells you will find 100 multiples of the entered number. Column B displays the numbers 1 through 100 and thus forms a multiplication table of sorts. You can enter any number of your choice in cell B10 also, so that you are not limited to the first 100 multiples of a number. You can see any multiple of any number!

Exercises

1. Enter 4 and 22 and observe their multiples. How many are even, and how many are odd? Why?
2. Enter 5 and 23 and observe their multiples. How many are even, and how many are odd? Why?
3. Enter 6 and 12. Are there some numbers in common in the lists? Do you notice a pattern? What about 4 and 8, 7 and 14, and 35 and 70? Do you see the same pattern?
4. Enter 2 and 6 and observe whether there are common multiples in the two lists. Try 7 and 21, 4 and 12, 9 and 27. How do you account for the frequency of the common multiples?
5. From your observations above, what would you predict to see in the way of common multiples in the lists for 9 and 45, 2 and 18, and 6 and 24? Check them out.
6. Look at the lists for 24 and 18. What is the lowest number you can find common to both lists? (It's called the *lowest common multiple*.)
7. What is the lowest common multiple of 60 and 40; 15 and 20; 4 and 7; 100 and 130?
8. What is the 157th multiple of 19; the 4985th multiple of 61?
9. MAKE UP YOUR OWN EXERCISES!

Basic Arithmetic

Although it is necessary for students to develop a degree of proficiency with arithmetic, it remains a fact (unsettling to some) that computational devices are used in the real world today to do the arithmetic that people used to do by hand. That's life right now! Spreadsheet models can let students see what is really going on. When a student is struggling to find an answer using pencil and paper, the answer is often his or her *only* concern. The beauty of the process, the logic of the process, and the relationships among many answers never become obvious.

Several spreadsheets in this collection simply calculate for us. They do the work for the student. If the teacher sees that they are used by keenly observant students, we believe these sheets will be of great benefit.

Open DOING THOSE OPERATIONS from the Basic Arithmetic folder.

When you enter two numbers of your choice in cells G5 and G6, all four basic arithmetic operations are performed simultaneously below. If handheld calculators are an acceptable tool for you, then this spreadsheet should be, too.

Exercises

1. Find the sum, difference, product, and quotient of 80 and 16. Change the order and see which operations are affected by order and which are not. Use different numbers and try again.

2. Find number pairs whose products are greater than their sums. Now find number pairs whose sums are greater than their products. Find number pairs whose sums and products are equal. Can you make some exact statements about when each situation occurs?
3. Consider pairs of even numbers. Are sums, differences, products, and quotients (if divisible) even? Study odd numbers, prime numbers, and so on.
4. Study positive and negative numbers and deduce the *sign rules*.
5. Divide by 0. What does the spreadsheet do? Why?
6. MAKE UP YOUR OWN EXERCISES!
7. *Extension:* Create your own spreadsheet calculator that incorporates raising to powers. (The symbol ^ means "raise to a power." 6^2 would be 36. Be aware that exponents yield big results quickly. You will frequently overload a cell's capacity if you try large numbers.)

Open AVERAGES from the Basic Arithmetic folder.

We find averages by adding and dividing. Often, we don't bother to look at the cause and effect involved with increasing and decreasing numbers in the list. Although it is important to find an average, it is equally important to be able to judge the changes necessary to bring about a given resulting average.

This spreadsheet, using adjacent columns, lets the student experiment with the averages of two numbers through the averages of ten numbers simply by changing the numbers in the columns beginning with row 8. Instant feedback on the average is provided at the bottom of each column.

Exercises

1. Find the average of 57, 35, 86, 33, 8, and 27.
2. In the column for the average of five numbers, give yourself test scores of 90, 88, 91, 85, and 0. (You forgot to study for that last one.) Now move to the next column (for six numbers) and enter the same five test scores along with a 100. How much did your average improve? Does a 0 in a list of test scores make a big or a small difference to you?
3. Perform similar experiments with fewer test scores and with more test scores to determine when a 0 would have a greater effect and when it would have a lesser effect.
4. Go to the average-of-3 column and enter any three numbers. Note the average. Change any one of the numbers to make the original average increase by exactly 1; by exactly 2; by exactly 5. Can you learn to predict what change it would take in one number to increase the average by 4? By 6? By 10?
5. Perform similar experiments in other columns. Can you come up with a rule for changing the average of any number of numbers by any amount?
6. MAKE UP YOUR OWN EXERCISES!

INSTRUCTIONAL MODELS AND IDEAS

Open PROPORTIONS from the Basic Arithmetic folder.

"This is to this as that is to that." Proportions occur in so many places that knowing how to solve them is a very important skill. It is also important to have a *feel* for them, and this spreadsheet shows proportions in action.

Notice that you must scroll up and down to find all four situations. In each situation, the unknown is designated by the letter x. A student can replace the three accompanying terms with numbers of choice. To the right of the proportion will be the solution. The operating principle is the same no matter which term you are trying to find. Just locate the situation that applies.

Exercises

1. Solve the proportion $2/3 = 60/x$.
2. Solve the proportion $x/28 = 120/160$.
3. Solve the proportion $13/8 = x/56$.
4. Bill's computer printer produced 12 graphics pages in ten minutes. How long will it take for that same printer to produce 30 similar pages?
5. Pick any case and alter the given entries according to a pattern. Observe the patterns in the revealed unknown.
6. MAKE UP YOUR OWN EXERCISES!

Open RANDOM NO. GRID from the Basic Arithmetic folder.

This spreadsheet provides the fodder for practice in basic arithmetic processes. It creates a 15 × 15 grid of random numbers in any range you specify, making a form on which to "play" with positives and negatives and odds and evens and sums and differences and products and divisibility and primes and proportions....

Designate the range of numbers by making the appropriate entries in cells G6 and G7, then scroll the screen so that only the grid shows. Preferably, print the sheet on paper so that you can use it as a handout for class activity. You might print different grids for different students or teams of students.

Exercises

1. Find the greatest column or row sum.
2. Find the greatest column or row product.
3. Alternately add and subtract each number throughout the grid.
4. Find all three-number combinations (horizontal, vertical, diagonal, L-shaped) in which two numbers add up to the third.
5. Find all three-number combinations in which one number is the product of the other two.
6. Find the longest chain of even (odd) numbers you can.
7. How many different prime numbers can you find in the grid?

8. How many 3 × 3 grids can you find in which the two diagonals have the same sum? The same product?
9. How many three-number combinations can you find where the sum of two numbers equals the third on a twelve-hour clockface—for example, 9 + 9 = 6, 1 + 2 = 3, 8 + 5 = 1? (Award one point for every "nonwraparound" example, such as 1 + 2 = 3, and two points for each "wraparound" example, such as 8 + 5 = 1.)
10. Repeat the exercise above with an eleven-hour clock, or any other clock.
11. Find every combination of three numbers where one number is the remainder when a second is divided by a third.
12. Find every four-number combination that represents, in order, a proportion, like 3 6 4 8 (3/6 = 4/8).
13. Find every three-number combination where one number is the greatest common factor of the other two.
14. Find every three-number combination where one number is the lowest common multiple of the other two.
15. Choose any low number in the leftmost column and any high number in the rightmost column and find a continually ascending path between them, left to right.
16. Circle the largest (smallest) three-digit number you can find; four-digit number; twelve-digit number.
17. Using many calculated sheets as samples, estimate the probability of primes (evens, odds) in a particular range of numbers.
18. Make up a bingo game.
19. Make up any board game.
20. Find as many buried equations as you can; for example, 5 4 2 7 in a consecutive combination could be 5 + 4 − 2 = 7. Award points on the basis of the length and complexity of the equation.
21. Assign a denomination of money to each column. Calculate the amount of money in each row.
22. Can you find any palindromes (numbers that read the same forward and backward)?
23. Circle all multiples of a given number.
24. Circle all divisors of a given number.
25. Find means, medians, and modes.
26. MAKE UP YOUR OWN EXERCISES!

Open EXPANDED NOTATION from the Basic Arithmetic folder.

The understanding of place value is essential to every move we make in the study of arithmetic. This sheet reinforces the paper-and-pencil activities that students perform. Allowing students to *observe* the expansion makes a different kind of impact from what results if they struggle with the expansion themselves.

Exercises

1. Have the sheet expand 5384.297.
2. Have the sheet expand 7050.06.
3. Use the digits 6, 3, 2, 8, and 5 to create the smallest number this sheet can form; the largest number this sheet can form.
4. MAKE UP YOUR OWN EXERCISES!
5. *Extension:* Do you dare try this for a different base?

Open DECIMAL EQUIVALENTS from the Basic Arithmetic folder.

This spreadsheet helps in comparing fractions. Decimal equivalents of fractions, rounded to four different precisions, together with pie graphs representing the given fractions, provide two quick methods of comparison. The graphs particularly give a good feeling for just how much bigger or smaller one fraction is than another.

Any fraction greater than 1 will still be correctly represented by its decimal equivalents, but the pie graphs will fail.

Exercises

1. What is the decimal equivalent for 3/8?
2. How do the decimal equivalents for 4/7 and 13/19 compare?
3. Order the fractions 41/52, 4/5, and 79/99 in descending order.
4. True or false: Given any fraction, you get an equivalent fraction by adding the same number to the numerator and denominator? For instance, is 7/10 equivalent to (7 + 2)/(10 + 2) or (7 + 7)/(10 + 7)?
5. True or false: Given any fraction, you get an equivalent fraction by multiplying the numerator and denominator by the same number? For instance, is 5/6 equivalent to 5*4/6*4 or 5*100/6*100?
6. Start with 10/10 and its solid-red circle graph. Decrease the numerator successively by 1—making 9/10, 8/10, 7/10, ..., all the way to 0/10. Notice the way the graph changes. Now start with 1/1 and increase the denominator successively by 1—making 1/2, 1/3, 1/4, ..., all the way to 1/10 (or beyond). Does the graph change in the same way—at the same pace? Can you explain your observations?
7. Put 3/4 into the first fraction's position and 1/10 into the second. Observing the graphs, can you visualize what the graph should look like for the sum of 3/4 and 1/10? Find that sum and enter it into the third fraction's position. Does its graph look like your prediction?
8. MAKE UP YOUR OWN EXERCISES!
9. *Extension:* Can you make a sheet that shows the sum of the three fractions along with its decimal equivalent and graph? (Hint: You might want to look over at cells L7 and L8.)

Geometry

The study of geometry is a study of a beautiful mixture of the qualitative and the quantitative. It is at once a study that thrives on both the description of space and the numbers that lend order to that space.

Here are three spreadsheets that address topics in geometry. Think of how many more you could produce!

Open RECTANGLE AREA VS. PERIMETER from the Geometry folder.

Linear units versus square units and the dynamics of the rectangle can be investigated with this spreadsheet. Two rectangles, the red rectangle number 1 and the gray rectangle number 2, provide the opportunity to alter lengths and widths. The sheet provides areas and perimeters numerically along with a graphical comparison.

Exercises

1. If a rectangle has a length of 60 and width of 32, what are its perimeter and area?
2. If two rectangles have equal widths but one length is three times the length of the other, how do the perimeters and areas compare?
3. Can you provide dimensions for the two rectangles so that the red perimeter is less than the gray perimeter but the red area is greater than the gray area?
4. Can you make two rectangles of equal perimeters but unequal areas?
5. Can you make two rectangles of equal areas but unequal perimeters?
6. MAKE UP YOUR OWN EXERCISES!

Open PLAYGROUND DESIGN HELPER from the Geometry folder.

This sheet simply provides a calculator for the area and perimeter of a rectangle given its length and width; the area, perimeter, and hypotenuse of a right triangle given its legs; and the diameter, area, and circumference of a circular sector given its radius and measure of arc in degrees. It can support quantitative reasoning in a variety of situations involving those figures.

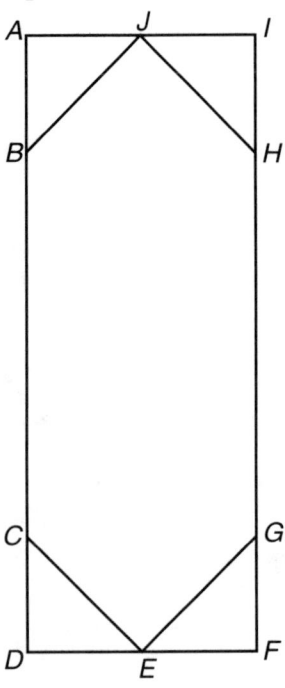

Exercises

1. *AIFD* is a rectangle where *DF* = 15 and *AD* = 40. *J* and *E* are midpoints. Triangles *ABJ*, *HIJ*, *CDE*, and *FGE* are isosceles.

   What is the perimeter of figure *JBCEGH* and what is the area of the same figure?

INSTRUCTIONAL MODELS AND IDEAS

2. *AIFD* is a rectangle where *DF* = 24 and *AD* = 56. Diagonals *AF* and *DI* are drawn. The circle, tangent at points *B* and *C*, is centered at the intersection on the diagonals.

   What are the perimeter and area of the figure bounded by segments *BA*, *AI*, *IC*, and arc *CEHB*?

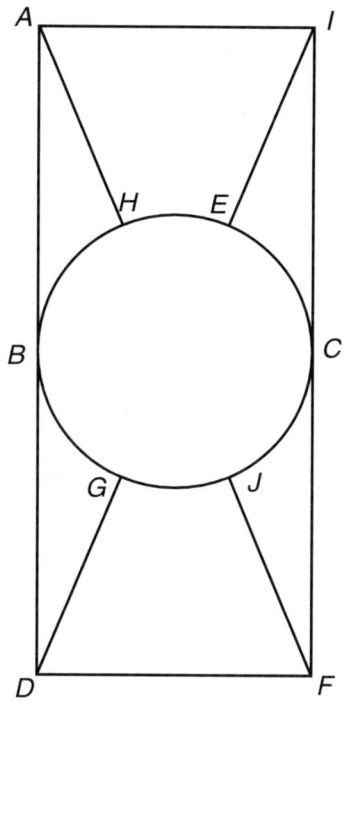

3. *E* is the midpoint of leg *AC* of isosceles right triangle *ABC*, and *CB* = 22.

   What are the perimeter and area of the figure formed by segments *FB* and *BG* and circular arcs *GE* and *EF*?

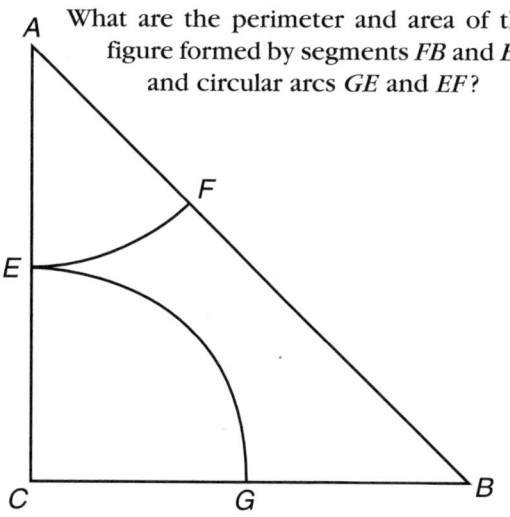

4. What is the area of your checkerboard showing when the checkers are in place ready to begin the game?
5. Design a new playground with a drawing program and calculate the areas and perimeters of the playground.
6. MAKE UP YOUR OWN EXERCISES!
7. *Extension:* Expand the spreadsheet to address the volumes of three-dimensional objects in a space.

Open GENERATING PI from the Geometry folder.

Pi, that famous old number without which the study of circles (and lots of other things) could not proceed, has a variety of derivations. One involves an imaginary square dart board centered at the origin of a coordinate plane. The square has a side of 2 and has inscribed in it a circle of radius 1.

This derivation of pi goes as follows: I throw darts randomly at the square board. Some of the darts fall inside the circle, but every dart hits the square board. I never miss the square board. The ratio of the number of darts that hit the circle's interior to the number of darts that land in the square (both in and out of the circle) approximates the ratio of the area of the circle to the area of the entire square.

That is, this proportion holds true:

$$\frac{\text{number of darts in the circle}}{\text{number of darts in the square}} = \frac{\pi r^2}{s^2}$$

Or, using $r = 1$ and $s = 2$, we find that

$$\frac{\text{number of darts in the circle}}{\text{number of darts in the square}} = \frac{\pi}{4}.$$

By cross-multiplication,

$\pi = 4 *$ (number of darts in the circle/number of darts in the square).

This sheet simulates throwing darts, checks to see which darts land in the circle and which darts land in the square but miss the circle, and calculates the value of pi as noted above. See cell G20 for that result. We've arbitrarily used 200 throws of the dart in this investigation.

By recalculating the spreadsheet, either by choosing Calculate Now from the Calculate menu or by pressing $\boxed{\text{Shift}}$ $\boxed{\text{Command}}$ $\boxed{=}$, you can see new estimates for pi generated.

Exercises

1. Use a separate spreadsheet to total and average the pi approximations you observe in cell G20. How many approximations do you need before your average "settles in" at 3.14?
2. Extend this pi generation sheet to 400 dart throws and reconsider question 1.
3. *Extension:* Can you make another spreadsheet, using similar techniques, to estimate the area of the square formed by connecting the midpoints of the sides of a 2×2 square?

Probability

The world of chance is exciting. Mathematics is alive and well in that world. Unfortunately, we often haven't gotten around to teaching probability. With several excellent commercial programs now available, there is little reason for not being able to experiment with coin tossing, dice rolling, marble drawing, dart throwing, and other introductory probability activities. Although the spreadsheet cannot compete dramatically with those programs and their color and animation, it does provide the structure and facility to design your own experiments with your own particular twists. It's worth our while to take a look at a few examples.

Author's note: Those of you familiar with the Apple II version of this book may remember that we frequently used a countertechnique for accumulating the number of trials and incrementing totals. For instance, were we counting how many trials in cell E3, we would enter the formula =E3+1 in that cell. Unfortunately, the more modern spreadsheets such as ClarisWorks don't allow that maneuver, calling it a "circular reference." The same problem would arise in ClarisWorks were we to try counting doubles in dice rolling by entering =IF(C1 = C2, E3 + 1, E3). We therefore compromise here by being content with a fixed number of trials, say twenty.

A number of ClarisWorks "gurus" have been consulted, but none has suggested a way to bring "progress" up-to-date. Any readers—especially students—who have a constructive suggestion are welcome to submit it.

Open TEN-COIN EXPERIMENT from the Probability folder.

This spreadsheet shows ten coins being flipped at once twenty times. Columns L and M record the number of heads and tails in each of those twenty trials, whereas columns N and O provide the cumulative averages for heads and tails, respectively. The class should discuss the subject of how probability experiments profit from the cumulative effect.

To make a new set of twenty trials, either go to the Calculate menu and select Calculate Now, or better yet, use the keyboard equivalent of [Command] [Shift] [=]. Every time you use [Command] [Shift] [=], you should see entirely new data.

Exercises

1. Does the so-called "50-50" coin fall five heads, five tails very often? Is there a combination that falls more often? Recalculate the sheet several times. Do you answer the same each time?
2. What seems to be occurring in columns N and O? What does the word *cumulative* mean here?
3. Select cells N3 through O23 and draw a line graph. Click back in the spreadsheet and recalculate the sheet several times. What do you notice? Explain what's happening.
4. MAKE UP YOUR OWN EXERCISES!
5. *Extension:* Unlock all cells and extend the experiment from 20 trials to 100, 200, or 400 trials. Reconsider questions 1-3.
6. *Extension:* Can you make the coin an unfair coin? For instance, in cell B4 change the function to RAND(3), then test for RAND(3) <=2, to make a heads result twice as likely, for instance. Reconsider questions 1-3.

Open FIVE-DICE EXPERIMENT from the Probability folder.

Row 3 of this spreadsheet invites the student to enter the number of faces to be on each of five dice. Of course 6 is the most common number of faces. Rows 6 through 25 show the results of twenty rolls of those five dice. Columns H and I provide the means for the individual rolls and cumulative rolls, respectively. In rows 29 and beyond can be found the number of times individual 1's, 2's, 3's, ..., are rolled.

Exercises

1. What is the greatest total that a set of five 6-faced dice can give you? Does it happen very often? Recalculate the sheet using ⌈Control⌉ ⌈Shift⌉ ⌈=⌉ several times.
2. What is the least total that a set of five 6-faced dice can give you? Does it happen very often? Recalculate several times.
3. What do you expect as the mean (average) roll of five 6-faced dice? Explain. Does column H support that? Do you see support for your answer in column I?
4. Change the number of faces in cell G3 to 15. Reconsider questions 1–3.
5. MAKE UP YOUR OWN EXERCISES!
6. *Extension:* Redesign the spreadsheet to use more than five dice or to have more than twenty trials.

Open URN WITH REPLACEMENT from the Probability folder.

Here we provide a virtual urn and an unlimited number of green, red, and yellow marbles to put into the urn. The sheet tells us the probability of drawing one of those marbles based on the number of each entered. Further, it lets us (1) perform twenty trials of the drawing of three marbles with replacement after each draw and (2) receive a report on how many times all marbles are the same and how many times all marbles are different.

In rows 36 through 38, the spreadsheet counts the individual numbers of green, red, and yellow marbles. Those data, in turn, contribute to decimal representations for each number drawn back up in cells G7, G8, and G9. Thus, the a priori and experimental probabilities can be compared.

Exercises

1. Enter fifty marbles of each color. Before running the experiment, observe the probabilities. Using ⌈Control⌉ ⌈Shift⌉ ⌈=⌉ a few times, describe how the experimental probabilities compare with the predicted probabilities.
2. Now enter one marble of each color. Reconsider the tasks in question 1. Are your observations the same?
3. If you were given the task to maximize the number of "All the same," how would you choose to enter the number of each color of marble? Try it. Refine your strategy if needed. Describe what advice you would give to others to accomplish a lot of "All the sames."
4. Reconsider question 3, except for "All different."
5. MAKE UP YOUR OWN EXERCISES!
6. *Extension:* Extend the number of trials to 400 to see whether results are more clear-cut.
7. *Extension:* Can you change the sheet to perform the same experiment, but without replacement?

Open EIGHT-COIN BINOMIAL from the Probability folder.

This title indicates that we are going to toss eight coins and perform the classic binomial experiment. Each coin can come up either heads or tails, so the eight coins allow for nine different possible results—8 heads and 0 tails, 7 heads and 1 tail, and so on, all the way through 0 heads and 8 tails. The yellow section tracks this information for 200 trials at a time.

$\boxed{\text{Command}}$ $\boxed{\text{Shift}}$ $\boxed{=}$ is the key combination to retoss the coins. Highlighting the yellow section and then calling for a bar graph yields a revealing, and familiar, look at the results. It's useful to leave the graph in place on the screen, click back in any cell of the spreadsheet, and then retoss the coins. The graph changes before your eyes.

Exercises

1. Discuss why the results in the middle of the list are bigger than the results at either end of the list.

2. *Extension:* Extend the sheet to see 400 sets of coins tossed.

3. *Extension:* Change each of the many coins from a fair, "50-50" coin to a coin that is twice as likely to come up heads as tails. Describe the new look of the bar graphs you see and explain why you're seeing those results.

Open MATCHING BIRTHDAYS from the Probability folder.

Two people walk into your room who are *random* people. What is the likelihood that they have the same birthday?

Chances are that the younger the child, the more he thinks that the likelihood is high. After all, he has a sister whose birthday is the same as the president's, or Mrs. Miller across the street has a birthday that coincides with the day his brood of hamsters was born. Of course, scanning your collection of nonrandom people and looking for matching birthdays that you have already heard of is not the same question, but that won't impress young students.

This spreadsheet looks at 200 pairs of people we have never met. It judges whether they have the same birthday and counts, in cell F4, the number of matches.

This spreadsheet uses 1997, a non-leap year.

Exercises

1. Look at the first date in cell A4. How many chances are there that the date in cell B4 matches it? (1 in 365)

2. In order to have a "good chance" at one total match, how far should the spreadsheet be extended downward? Do that, and be sure to change the information in cells F3 and F4.

3. *Extension:* Create a spreadsheet that estimates the probability of any two of three people being born on the same day.

Communication across the Curriculum

We ask students to write in a wide variety of contexts. In social studies, students write reports on historical events or the states of the Union. In language arts, writing book reports and stories is a common practice. In science, we might ask students to describe an experiment or a natural process. It's only fairly recently that communication using the language of mathematics has been pushed front and center.

When personal spreadsheet programs first came on the scene the better part of two decades ago, they were aimed exclusively at members of the business world. They were not particularly friendly to use, and they featured functions that only real estate agents and tax accountants could love.

Modern spreadsheets still offer much to those types of users, of course, but in this section we suggest a very different classroom application. Here we use two languages essential to a large percentage of our world's young people—English and mathematics—and bring them together in perhaps some unexpected ways to address cross-curricular tasks.

Open MY STATE REPORT from the Communications folder.

It seems that all U.S. students sometime relatively early in their academic careers are faced with the preparation of a report on a state. They use library resources to determine the capital city, the major products, the population and area, and any of a large list of facts about the state, then put them into narrative form. They add a few pictures, place the work in a folder, and the report is in final form.

Consider having students create a state report spreadsheet template, perhaps individually, perhaps in teams, or perhaps as a whole group. In this spreadsheet, we have entered a list of data in straightforward list form. Some of the data are expressed as labels; some are expressed as values.

Below the line, in paragraph form, the student writes about the state. In this example, the red entries are the result of formulas and functions. Click on those red entries and view the formulas and functions in the entry window. Some formulas are very simple—just copies of entries from the list above. Other formulas use simple arithmetic, and still others might use conditionals, such as where the word *some* describes the land below sea level.

The information here is for New York. Another student could enter the corresponding information for Arkansas in the list above and the entire narrative would change in response to those entries.

There is no limit to the information that can be incorporated in the report, and there is no limit to the way that information can be manipulated within the narrative. Research, sentence structure, fiction and nonfiction writing techniques, mathematical reasoning, vocabulary, science process—spreadsheets can encourage the accomplishment of a number of curricular outcomes at once.

Reports on states are just one of many ideas. Science reports, mystery stories, autobiographies, and a profile of the class are some others.

Exercises

1. Write a spreadsheet report about our solar system.
2. Write a spreadsheet version of your favorite nursery rhyme.
3. Write a spreadsheet description of the animals on your farm.
4. Write a spreadsheet description of your classroom, then pass along the template to other classes.
5. Write a spreadsheet news article about things that happened in the school office yesterday.
6. Write a sports statistics report on your favorite baseball (football, basketball, hockey) player, updating the data after every game.
7. MAKE UP YOUR OWN EXERCISES!

Miscellaneous

There are always those spreadsheets that don't fit neatly into the usual categories. They can be the most fun, at least when the teacher is not looking. These examples are included here "for free."

Open MESSAGE FROM SPACE from the Miscellaneous folder.

You may enter any "message from space" in the form of X's and spaces into the pink square area in rows 5 through 10. It is that configuration that then must be interpreted by the supersecret computers deep underground in central Australia—computers that by good fortune we've been able to tap into with our ClarisWorks spreadsheet.

As you scroll down the sheet, you will see twenty numbered grids with their own configurations of X's and dots, the dots standing for spaces because they're easier to see than spaces. What are the rules that apply to each of these areas?

Let's consider how we get from the original pink square to the first green one. For each individual cell in the pink square, the computer counts the X's in adjacent cells. Adjacent cells are those immediately to the left or right of, above, or below the given cell as well as any corner-to-corner touching cells. If the number of X's counted is greater than or equal to half the total number of adjacent cells, an X is put into the corresponding cell in the green square. Otherwise, a dot is put there.

The same rule holds for the development of any new square as you read down the page.

This spreadsheet is set for manual calculation. That is, once you complete the entries in the pink square, then you call for all the green squares to take on their new look by accessing Calculate Now from the Calculate menu, or by using the [Command] [Shift] [=] keyboard equivalent.

One more note: Take care to use uppercase X's. They are easier to see, and the rule recognizes them only.

Exercises

1. Enter the configuration at the right in the pink square.

    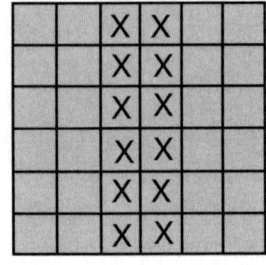

    You should notice that all the green squares have the very same configuration of twelve X's as your original pink square. Take two cells, one with an X and one with a space, and explain why that happened.

2. Remove any single X from a cell in the original pink square and observe the results. If you removed an X from any cell but a top or bottom one, all green squares had a resulting count of twelve X's. If you removed an X from a top or bottom one, you had a different result. Explain why there is that difference.

3. Starting with the original pink configuration, can you remove any four X's and have the resulting green squares remain at a count of eight? How about removing any six X's and maintaining a count of six? If you did that for one through twelve, what would you observe? (Patience is a virtue, and often very necessary in research.)

4. What is the lowest number of X's you can start with and have the population build up consistently? What is the configuration?

5. MAKE UP YOUR OWN EXERCISES!

6. *Extension:* Enhance the spreadsheet by including a cumulative average (mean) for each green square.

7. *Extension:* Change the rules so that the count that determines whether or not an X appears includes the cell itself along with its adjacent neighbors.

8. *Extension:* Research the Game of Life and establish those rules in this spreadsheet. (You can easily find a number of freeware or shareware versions on the World Wide Web.)

Open WORD VALUE from the Miscellaneous folder.

If you assign each letter the number equal to its position in the alphabet, any word you type into cell E2 will be evaluated with its total letter value. Arbitrarily we impose a twenty-letter limit on the word.

A number of challenges linking numbers to vocabulary can be investigated with the use of this spreadsheet.

Exercises

1. Find the mathematics term with the greatest possible total value. Be prepared to define the word you claim.

2. How many words can you find in which each letter represents a prime number?

3. How many words can you find in which each letter represents an even (odd) number? What is the longest such word you can find?

4. Can you find a word whose value list represents an arithmetic sequence?

5. Highlight from cell E5 through G5 and down through the value of the last letter of your word. Make a line graph of those results. Cut that graph to the Clipboard. Try to find another word whose graph never would intersect the first. (Create your second graph and paste the first one back for comparison.)

6. If you're a musician, use the MOD function to help you assign each value to an eight-tone scale. Then "sing" each word. Can you find a word that "sounds like" any familiar song? Can you put harmony to such a "word melody"?

Word Problems

Word problems in arithmetic and algebra books are notoriously of a cookbook nature—the big difference being that Tony has the coin collection in one problem and Mary has one in the next problem. When you turn to the page of distance problems, planes leave an airport and fly in opposite directions, then trains part company from a common city. It's not that finding missing quantities in these contexts is not important but that the canned, memorized approaches to the solution of these problems is suspect.

A spreadsheet can be made to generalize these word-problem types to set up variable categories and functions to calculate coin-collection values or distances from an origin. With such models, students can engage in estimation or trial and error and be more immediately involved in bringing a solution to the problem by matching the facts of the problem to what occurs on the screen.

Open MANAGING MY MONEY from the Word Problems folder.

By entering the number of fifty-dollar bills through pennies in the indicated column, the student can view the number of pieces of money, the accumulated values of each denomination, and the total value of the collection. The bank is totally flexible in matching the conditions of any typical money problem from an arithmetic or introductory algebra text.

Exercises

1. *a)* Make $13.42 using the greatest number of pieces of money.

    b) Make $13.42 using the smallest number of pieces of money.

    c) Make $13.42 using exactly 19 pieces of money.

    d) Make the best change combination for $13.42 if you were given $20.00. *Hint:* Use the count-back method.

2. Sarah's bank contains a total of $37.20. She has twice as many dimes as five-dollar bills and two more nickels than dimes. She has no other denominations. How many five-dollar bills, dimes, and nickels does Sarah have?

   Write about your strategy for answering this question. Explain how you knew you had the correct combination of five-dollar bills, dimes, and nickels.

3. A coin collection has a total worth of $9.48. It is composed of only quarters, dimes, nickels, and pennies, and the number of each is a multiple of 6. How many of each kind of coin is in the collection?

4. You have some of each of five different denominations and no other money. You can choose to have 1 of one denomination, 2 of another, 3 of another, 4 of another, and 5 of another. Where would you put the 1, 2, 3, 4, and 5 to have the greatest possible total money and where would you put them in order to have the second greatest possible total money?

   Explain in writing your thinking. Describe your strategy if the question asked for the least possible total money and the second least possible total money.

5. MAKE UP YOUR OWN EXERCISES!

6. *Extension:* You've studied the following Latin American countries: _____, _____, and _____. Make a spreadsheet similar to this one but for the monetary system of _____.

Graphing

Many wonderful graphing programs are available that represent relations and functions quickly and accurately. From graphs thus produced the properties of parabolas and ellipses can be investigated. Solutions of systems can be identified. This spreadsheet's graphing capabilities fall way short of those of many other products on the market for those purposes; however, once again the process of developing a useful spreadsheet/graphing combination is important. Here are some examples of the kinds of things that might be done.

Open PARABOLA from the Graphing folder.

Row 6 of this spreadsheet displays a common representation of an equation of a parabola, and row 7 invites the entry of coefficients in C7, F7, and I7. One further entry from the user will be accepted in B10, where the leftmost x-value can be specified.

After making any changes in those four cells, the graph should adjust to show the parabola over the domain you've indicated. The scale on the axes needs to be taken into consideration. Scale changes happen at the whim of the program and certainly can furnish some interesting twists in your view of the graph.

Exercises

1. Produce the graph of $y = 2x^2 + 4x - 6$ beginning at $x = -10$.

2. Produce the graph of $y = -3x^2 + x + 1$ beginning at $x = -4$.

3. Estimate the axis of symmetry of $y = x^2 + 10x - 5$.
4. Estimate the coordinates of the minimum or maximum point of $y = -4x^2 - 4x + 2$.
5. MAKE UP YOUR OWN EXERCISES!
6. *Extension:* Incorporate into this spreadsheet the calculations for the axis of symmetry equation and maximum and minimum points so that you can check your estimates against reality.

Open SOLVE LINEAR SYSTEM from the Graphing folder.

Like the PARABOLA spreadsheet, this one suffers from capricious scales on the axes. But as you try solving a system or two with it, you may have some ideas for refinement or some new creations that will help you with problem-solving tasks you have. (Yes, graphing calculators are probably much more efficient for the serious investigation of linear systems!)

Note that you can control the domain for each function separately in cells B13 and D13.

Exercises

1. Estimate the solution to the system
$$4x - 5y = 0$$
$$-3x + y = -11.$$
2. Estimate the solution to the system
$$-x + 9y = 2$$
$$x + y = 8.$$
3. MAKE UP YOUR OWN EXERCISES!

Open COORDINATE PLANE from the Graphing folder.

Here's a spreadsheet that is very close to being frivolous! By entering the x- and y-coordinates of a point in cells R1 and U1, respectively, you cause the letter X to locate itself at the designated location on the coordinate plane represented by the spreadsheet. (Talk about the proverbial peanut and the sledgehammer!)

Although we don't recommend using this spreadsheet to practice locating points, it does, we hope, suggest some more interesting refinements.

Exercises

1. *Extension:* Refine this spreadsheet so that it shows the graphs of any equation of the form $x = a$ and that of $y = b$.
2. *Extension:* Refine this spreadsheet so that it shows the graphs of $x > a$ and $y > b$ (or $<, \leq,$ or $\geq$).